Facebook Ma

Tips

Zero Cost Facebook Marketing Plan for Small Business

Table of Contents

Introduction

Congratulations on downloading **Facebook Marketing Strategies: Zero Cost Facebook Marketing Plan for Small Business** and thank you for doing so. The world of Facebook marketing is growing increasingly chaotic and downloading this book is the first step you can take towards actually doing something about it. The first step is also always the easiest which is why the information you find in the following chapters is so important to take to heart as they are not concepts that can be put into action immediately. If you file them away for when they are really needed, then when the time comes to actually use them, you will be glad you did.

To that end, the following chapters will discuss the primary preparedness principals that you will need to consider if you ever hope to really be ready for a successful Facebook marketing campaign. This means you will want to consider the quality of your ads including the potential issues raised by their current costs, how they can be best utilized along with organic traffic and various reinforcements or fortifications you may need to have on hand in case of a new business opportunity.

With ads out of the way, you will then learn everything you need to know about groups and communities. Rounding out the three primary requirements for successful Facebook marketing, you will then learn about crucial page management principles and what they will mean for you.

There are plenty of books on this subject on the market, so thanks again for choosing this one! Every effort was made to ensure it is full of as much useful information as possible. Please enjoy!

Chapter 1: What Is Facebook Marketing?

Since Facebook stores data about its users which are entered spontaneously in their profile (e.g. age, location, interests), it has a very good idea of who its users are and what they are interested in. For example, just put a like on a football page to make them understand that you are following this sport.

You will understand then that the potential of subjecting these well-defined users to targeted advertisements is enormous.

In the past years, social media has become one of the most effective advertising channels, able to get new leads (contacts) and turn them into paying customers.

Facebook Ads works in both B2C and B2B and there are many cases that show an increase in results, even 5 times, after taking advantage of advertising on Facebook.

The growth of Facebook is steady; both in terms of new users and marketing opportunities, and the budgets dedicated to the AD on social media have doubled worldwide in the last 2 years, from $16 billion in 2014 to $31 billion in 2016. In 2018 analysts expect a further increase of 26.3%.

To all of this, add that the number of active users every day on Facebook was about 1.23 billion people in 2017 and on average each of these spends 50 minutes a day between Facebook, Instagram, and Messenger. With Facebook Ads, you can reach each of these users with highly customizable targets such as interests, demographics, locations, actions performed on a website, and much more.

Should I Then Do Facebook Advertising?

For how often this happens, the answer is "it depends". You have to ask yourself what your goals are and once the mechanisms of advertising on Facebook are understood, the answer will be obvious to you.

If you want to intercept an audience that might be interested in your product or service, surely Facebook Ads will be for you, allowing you to submit your ads to a specific target of users.

If you have traffic on your site or e-commerce, regardless of how it is obtained, you may want to recontact those users to repropose your ads.

Do you know that when you are looking for a product and after visiting a site, later when going to Facebook, you will find yourself the sponsored ads?

It's called retargeting. You will be able to create customized user lists based on their actions, such as visiting a particular page (article, product, landing page, shopping cart, etc.) and exploiting them for well-targeted ads.

It will be essential to have a very clear strategy for our Facebook Ads campaign. It will be appropriate to understand where the potential customer is located within our sales funnel and submit different ads depending on whether it is far or near the purchase.

We certainly cannot offer the same ad to a potential customer who does not know our brand and someone that already has our product in his chart. This is just to say that Facebook has great potential for small businesses, but needs to be used properly to get the most out of it.

Chapter 2: Disadvantages of Facebook Marketing

How much do Facebook Ads cost? How is the price established?

It is the most frequent question in this environment, but unfortunately, there is no answer other than "it depends".

On what does it depend?

From what you are advertising to who your target is, how many competitors there are and what your goals are, just to mention some aspects.

Before reaching the total costs of a Facebook campaign, it is important to understand how these costs are established. Unlike traditional advertising methods, Facebook does not have a fixed price for each placement, but follows a system of an auction among advertisers to get the ad published; this is because Facebook users can only see a limited number of ads per day.

There are various factors that determine which ad will be displayed, who will be shown and at what cost:

- Who you are targeting.

As we will see later, the possibilities of targeting are many and depending on the target you choose, you will compete with other advertisers.

- Your relevance score, engagement and click through rate.

Facebook assigns to each ad a score of relevance from 1 to 10 depending on the associated target and its response. The higher the interaction and the number of clicks, the higher your score will be and the more likely your ad will be displayed at lower costs.

- The timing.

The more competitive the sector is, the higher the costs will be. There will be periods of the year, such as during certain holidays, where costs could increase significantly. You can choose from various options, but which one is best will depend on your campaign objectives.

You can pay in various ways:

- Per impressions: pay for each display of your ad.

It is better to use this option when you want to reach as many people as possible, for example in a brand awareness campaign.

- Per click: pay only if someone clicks on your ad.

This option is preferable when we want to take users to click on a link that will take them to the site or to a specific landing page.

- Per action: pay only when a user performs a certain action, how to fill out an entry form, buy a product or visit a certain page.

With this option, Facebook will show ads to people who are more likely to take one of these actions.

So you will understand that answering the question "how much does it cost?" is complex. The important thing is to set a budget and manage it in the best way possible, reaching potential customers in a careful manner and keeping an eye on the really relevant parameters, such as cost per conversion. Not knowing what you will spend is one of the main disadvantages of Facebook marketing, but still, it is something that can be overcome easily.

AdWords vs. Facebook

This is a question that they often ask me. The answer is always "it depends".

We need to understand what the goals are that we want to achieve with our advertising campaign. It is often fundamental to combine both strategies, it all depends on the type of question: whether this is latent or conscious (or both).

If the goal is to make branding and then stimulate users who do not know us and may be interested (latent demand), the best choice is Facebook Ads, which will allow you, as we will see later, to reach potential customers. You can do this with various types of targeted campaigns and get leads.

Similarly, you can also take advantage of ads on the Google Display Network to reach potential customers by submitting your banners to specific placements.

If users are already looking for your product or service, the right approach is to use Google AdWords by creating ads on the Search Network. In this case, the user is already in a much lower part of the funnel, therefore more inclined to purchase as he is looking for your product/service.

Obviously, in most cases, we will find both types of demand and we will have to work on both platforms jointly.

The key thing is to understand where the user is inside our sales funnel and act accordingly; we will never tire of stressing it.

Going inside, it will be useful to retarget users who have shown interest in the product or service working with both Facebook Ads and AdWords through ads on the display network.

Attention: what if we intercept potential customers on Facebook through FB Ads and these then look for us on Google but we are not positioned in an organic way (without paying) for that keyword?

Simple, they will click on a competition link. The risk is, therefore, to practically advertise competitors. Understand well then how important it is, in the absence of organic positioning with SEO, that we must also have Google ads on the search network to cover some keywords as well.

The fact is that Facebook marketing is not so powerful if done alone, is something that some people see as a disadvantage.

Chapter 3: Facebook Marketing for Small Business

Small and medium-sized businesses can use Facebook marketing strategies with high margins of success. In fact, with more than 2 billion active users every month, it is impossible to remove the blue social from your web marketing plan.

What Could Be the Goals of Facebook Marketing Referring to SMEs?

Brand Awareness

Facebook is a very important tool that enables small and medium-sized enterprises (SMEs) to make their products and services known, while at the same time cultivating a very direct relationship with interested users.

If on the one hand your community, made up of people who already know your products, can follow us on the blue social, on the other side it is possible to reach people who do not know us, through spontaneous sharing, or through sponsored ones. The latter, through the creation of the right audience, allow reaching to new people potentially interested in our products.

Customer Care

Facebook is also one of the websites that best lend themselves to customer care, which is assistance to its customers. Indeed, given the announcement of future updates of the algorithm of the views of the News Feed, focus on customer care could also prove successful in terms of awarding the content posted.

Promotional content can still be valid, but using your own social page as a place to solve problems and perplexities of its users can be a key to use highly because it is able to trigger conversations between friends, debates, and an engagement appreciated by Facebook algorithms.

Direct Sales

Facebook can also be used to sell your products or services. Like an e-commerce site, the platform lends itself to the possibility of direct purchase from the page, with huge benefits for users. For small and medium-sized businesses, this opportunity is an important resource for saving resources that would otherwise have to be spent on the creation and management of an entire site.

Obviously it must be said that those who hold an important business cannot simply rely on the social network of Zuckerberg to market their products online, but it is a fact that not a month passes in which the Menlo Park team does not make availability of some new function that favors those who want to sell via the web.

How to Create a Facebook Marketing Strategy That Works?

Before starting to take action on Facebook, it is good for small and medium-sized businesses to devote time to creating a well-designed communication plan.

First of all, the goals of the strategies to be put in place must be defined.

Secondly, the public must be identified, that is the buyer, the typical customer, tracing a sort of identikit of its main characteristics such as age, place of residence, interests, level of education and more (see chapter 6 for more information about this topic).

As for the content, it will be good to dedicate only 20% of them to the promotion of your products or services so as not to tire the user with continuous offers and hype.

Finally, the tools for checking the results must not be forgotten, with the choice of the most appropriate metrics to follow in order to understand the effectiveness of the steps taken along the road to achieving the designated objectives. For example, if the setting up of a valid customer care campaign has been done, one of the ways to evaluate the effectiveness of the actions carried out is the analysis of the number and quality of comments received, rather than that of 'likes'.

Facebook for Small Businesses

Why should you use Facebook to market your small business?

Facebook is about to touch the ceiling of a billion users according to the latest official data released in July. The people who connect to this social network are 955 every month and 552 million every day; more than half a billion, even the number of monthly users who connect to Facebook with a mobile device - mobile phone, smartphone, tablet.

Facebook is becoming, for many, a major source of information more and more often, instead of connecting to the homepage of newspapers to see what is happening. We scroll the dashboard reading and commenting on the news linked by our friends. The 2011 CENSIS report on the American company shows that Facebook is used as a source of information by 26.8% of Americans, a percentage that grows to 61.5% in the age group of 14 to 29 years.

The quantity and nature of our relationships have been radically changed by the possibility of keeping in touch with people we do not see daily in person, but for the most diverse circumstances, we feel close. They were our friends in the past, we shared a travel experience or study, or we met online and subsequently met live.

This allows us to listen and exchange opinions, information, points of view, emotions; in this impetuous dynamic of conversations. This is an always open bar where people pass from one group to another participating in dozens of discussions. The companies suddenly find themselves "degraded" to one voice among others, which must gain attention thanks to the importance of what it says without the possibility of massively occupying the spaces of visibility. Furthermore, we must learn to speak "with" people, which mean, first of all, to listen and respond.

Should your company/association/electoral committee/theater company/excitement have a presence on Facebook? In the vast majority of cases, the answer is yes. Do not fall into the trap of thinking that Facebook is exclusively the realm of lazy people; often times it is a great way to "feel the pulse" of your stakeholders and can intercept your needs and opportunities that otherwise you would not have known.

In addition to your site where people go once in a while, or maybe they never go back; on the contrary, many of them open Facebook every day, several times a day; and, if they find what you publish to be interesting enough for them to click on "like" or leave a comment, this makes you visible on the bulletin board of their friends, not in the anonymous way of a flyer tucked in the mailbox, but with the social support of word of mouth.

An effective presence on Facebook can help you:

- Increase your visibility, spreading the posts of your blog, the videos you shoot, and the photos you take.

- Establish a more intense relationship with your customers, better knowing their needs and obtaining important feedback on what you do.

- Motivate and gratify your "super fans".

- Promote and share initiatives, special offers, and new products.

Facebook Marketing Ads for Small Businesses

Each Facebook campaign consists of 3 levels and it starts from the campaign level, which consists of one or more ad groups.

As you have just read, for each campaign you create, you will have to choose a goal. This is the real distinctive factor at the campaign level.

At the Ad Group level (Ad Set), you will have to choose the target, the available budget, the publication times, the offer and the placements (placements).

Going down the hierarchy, at the level of the announcements, you can set the type of announcement (image, video, carousel, etc.), all the texts, the call to action (action button) and the destination links.

As mentioned, the structure is hierarchical, so if you pause (or delete) for example a group of ads, the same thing will happen to all ads below that group.

The Definition of the Goals

Now that we understand the structure of a Facebook campaign and what are the parameters to be set for each level, we are ready to launch our first campaign.

The first question is, therefore "What is the goal to be achieved?"

Do you want to sell a certain product, because maybe you have an e-commerce store, want to create awareness or reputation, do you want to have leads or what?

Often, in a complete web marketing strategy, we will have to create different campaigns for the different phases of the purchasing process. We can then create different ads depending on whether the target user does not know our brand, or knows it but does not know our product/service, or for example, knows our product/service and may be interested in a commercial offer.

Facebook itself in the creation phase will propose you different objectives divided into 4 macro-categories; let's see them in detail one by one.

Brand Awareness

When to use it: in large-scale campaigns, when there is not a particular action that you want to take to the user. This goal will be more attractive to large companies that can afford to launch campaigns for pure branding. For smaller companies, however, almost every other objectives will give better and more significant results.

Reach

When to use it: similar to the brand awareness goal, the reach objective is functional to reach the maximum number of users to which the ad will show. With the introduction of the rules, Facebook now allows you to put a cap on the frequency with which the ad is shown to the same user; in this sense, the goal for reach becomes very useful when you have to work with a relatively small audience and you want everyone to view the ad.

Traffic

When to use it: when we want to take users to a website, or for example on a landing page. It is a very interesting goal when promoting content, such as a blog post.

Leads

When to use it: the lead ads greatly simplify the signup process from mobile devices. When someone clicks on the ad, a form opens with all personal contact information already pre-filled based on the information they share on Facebook, such as name, surname, phone and email address. This aspect makes the process really fast and within 2 clicks, one to open the ad and one to send the information.

The only problem with this type of objective is that often the email address used to sign up for Facebook several years ago is obsolete and has not been updated for too long. In this case, we would get a useless contact. As a result, it has been seen that better conversion campaigns perform that point to specific external landing pages with data to be filled out.

Another aspect to keep in mind is that lead ads do not allow you to include all the information you want in the offer, like on a landing page. Therefore, for campaigns that require a great deal of cognitive attention from the user, a campaign for conversions will be more successful.

That said, in any case, it is always better to do a test between the two approaches and see which performs better, because each case and sector can behave differently.

Chapter 4: How to Use Facebook Groups

A Facebook group is a micro-community within the largest community of the network that, focusing on specific themes, attracts people to the target.

To this, we add that the groups have recently had the blessing of Facebook after the hard blow to the business pages in the latest updates of the EdgeRank algorithm.

What happened to the Facebook algorithm?

The turning point came with the announcement of Zuckerberg and the subsequent official confirmation on the blog of Facebook news. In fact, the Newsroom informs users of an epochal change: less visibility on the pages in favor of a more personal communication.

More visibility for the posts of friends and family; but also explicitly mentioned by Zuckerberg, to the groups.

Unlike pages, in fact, groups have a greater predisposition to the generation of discussions and not to the simple unilateral posting of content. Precisely this content with little interaction is the one against which Facebook wants to fight. The decision comes after the progressive descent of the engagement on the posts that are over the years has characterized the Social Network of Menlo Park.

What does this mean?

It simply means that the posts on the groups will have greater visibility and the possibility to reach more easily your target audience. It seems to me, therefore, an excellent idea not to neglect them.

How to Find Facebook Groups in the Target

Let us imagine, of course, that you have already identified the niche or sector on which to base your communication strategy. The first of the problems you need to solve is to find groups to distribute your content on. I show you a couple of techniques you can use.

1. Facebook Searches

If you have an active Facebook account in America, you can use the search bar to find groups that meet your requirements. For example, if you look for a group that deals with social, you can write "social media" on the search bar and then apply the filter "groups" (Facebook will initially search among posts, people and so on).

If you want to get even smarter, I recommend you to use Facebook in English. You will open other very interesting possibilities! Facebook in English supports the "Graph Search", thanks to which it is possible to create more complex searches such as: "Groups joined by my friends who like social media". That is, look for groups to which were added by part of my friends who like social media.

Or you might still be looking for users who like certain pages or groups in an area of interest. For example: "Friends who like Wired". You can then go to peek at the groups in which your friends have entered, going directly to their profile (privacy permitting!).

2. Use Facebook Tips

Of one thing we can be extremely sure of is that Facebook knows us more than we think.

It is for this reason that it will be easy for them to suggest groups that might be interesting to you based on the past history of your likes and the groups in which you are already inside.

Just go to the Facebook page dedicated to groups. On the upper part of the drop-down menu, you can enter the "Discover" tab. Here you will find lots of suggestions divided by categories, including the "local" one to find geographically close groups.

This feature is available on both desktop and mobile. In the latter case, you will go to the main menu> Groups> Suggested.

I remind you that the specific features of the groups were recently incorporated into the main app of Facebook, while the app "Facebook Groups" retired in September 2017.

What to Do Before Posting Content in Groups

I already know that you're cheating to publish your content on the groups of your interest, but before starting to spam I suggest you follow some rules.

1. Always look carefully at group usage policies.

Any self-respecting group manager sets up rules in their own group to prevent the occurrence of phenomena such as SPAM, excesses of OT (Off Topic, i.e. ending off the arguments to which the group is dedicated), too much promotion, trolling and so on.

This means that there may be rules on the amount of content that can be submitted to the community, for example. Or that you can publish external links to Facebook only with the permission of the admin.

Policies are often indicated in the first post above, often attached with a "pin". Or they could be in the "file" section of the group. Study these rules well and behave accordingly.

2. Keep to general rules of good behavior.

Apart from the policies, which rightly dictate contents, times and rhythms, there are also rules of good common life that should always be respected. If you produce content in large quantities:

- Do not spam it all the time.

- Always check that it is in theme with the group, that they are of quality and that they can be useful and relevant.

- If possible, add a note to the article, a textual status that for example highlights a sentence of the article or that poses a question capable of generating a valuable discussion.

Remember, these are not only your contents! Subscribe to the groups, be in silent mode, and then spam your news as if there is no tomorrow, these will certainly irritate the admin that at some point they might even think to kick you out!

Even more so later in this historical moment in which the use statistics of the group have been made available to the administrators.

My advice is to always balance your content with content from other sources. And it would be desirable to participate in the discussions, in addition to the normal posting!

If you already produce your content and think that it is sufficient for your strategy, it means that you skipped the previous paragraph. Go read it!

For a content strategy, whether linked to the development of your business or to your personal branding, you need news and content about your industry.

I know I'm repetitive, but I'll write it again: you cannot just talk about yourself! To find valid content on the web, you can use many tools like the simplest monitoring software (I think of Google Alerts or Talkwalker) or news aggregator sites like Feedly.

In the era of the so-called "content shock", where our ability to absorb information is definitely lower than the amount of content produced, the figure of the creator becomes of primary importance. Skimming the contents and proposing the best to a community, therefore, becomes a primary role.

Additional Tips

Create a Network for Your Loyal Fans

This group is made up of the most loyal fans, people who believe in your company, in your product, and in your values. Brands like Canva, who has a loyal fan base, called these groups "community groups" or "ambassador groups".

The one with the loyal fans is the most important group you can create on Facebook. It is important to create a positive word of mouth on your products, your company, and your activities. Be sure to make your fans feel special. You can do it in various ways; by sending a shirt or by putting a comment on them.

The most important thing is to thank them, heartily and regularly. A written note (maybe by hand), is much more pleasant than a gift. Thank your fans for helping you grow your community and achieve your goals.

Here are 5 ways you can use these groups:

1. Get feedback on new products.

You have a new product to launch but first, you want to know how it would be accepted? This is the right place to test it.

2. Attract new members.

If your company has a membership program, this is a great way to keep in touch with them and attract new ones.

3. Answer the questions.

How many times have you seen a question on a group or an online forum and have you thought that your product could be the ideal solution? Of course, you could answer the question with your private profile, but you will be more successful by responding to the members of your group.

Answer often to questions and to avoid spam (especially if there are many members in the group), tag two or three experts in the response that can respond appropriately.

4. Recruit experts.

Many people like to feel expert and be taken into consideration. And many like to share their knowledge with other people. Experienced people can help new consumers get closer to your product and buy it. Please note that it may be necessary to create a dedicated expert group.

5. Share the company's achievements.

If the company wins a prize or is named in an important publication, let the whole world know about it. You can then call your friends, family, post the news on your private profile, but in this way, you will not go very far. If your company receives a prize, share the news with the fan group.

That's why Facebook exists.

Help Your Customers

The second type of Facebook groups is dedicated to acquired customers. To build this community, invite people to join the group as soon as they make a purchase. Make customers aware of the existence of the group and that you would like them if they too registered.

Explain how the group works in the "information" section, as Golden Tote does. For example, members can get to know each other, share ideas and strategies and help each other in any way.

Be sure to send an email with the link to subscribe to the group and monitor who signs up or not. Do not forget to invite people who have not yet registered several times.

Ways to develop your group on Facebook:

- Be generous and promote a "give" policy.

For example, offer tips or tricks to use the less known features of your product and invite other users to share their findings.

You could invite members of your fan group to share tips and tricks with them too.

- Be transparent.

Respond personally to customer criticism. For example, acknowledge the customer who makes you notice what is lacking in your product and responds that you will immediately fix it.

It is not necessary to promise discounts or promotions to the disgruntled customer, but it is important to thank him for the report, to apologize and to promise that he will be contacted when the problem is solved.

- Be present.

Answer questions and comments as soon as possible

If you cannot give an immediate answer, still let your customer know that you will respond quickly. From suggestions, tips, tricks. Encourage other users to share their findings.

- Be inclusive.

Offer group members promotions and offers that are not on the official page. Make your customers feel part of an exclusive club. If a social media user shows a particular interest in your product, he also invites you to join the online group.

Grow Target Segments (With Multiple Facebook Groups)

One of the most important goals of marketers is to develop a "typical customer", that is to identify their customers and understand what their doubts, problems and how your product can meet its needs.

If you have different types of customers, create more groups on Facebook. For example, you could create a group based on the language spoken by its users. Maybe not all customers will want to be considered experts and to answer questions from other users; you can then create smaller groups to meet the needs of each client.

Chapter 5: How to Start a Successful Facebook Group

Maybe you do not know, but creating a group on Facebook allows you to share photos, videos and anything else just with whoever you want. Thanks to groups, you can interact with a small circle of people without the others seeing what you're sharing with them.

Even if you do not consider yourself a social network expert, you can feel comfortable. In the next paragraphs, you will find all the information you need to create a Facebook group, whether you intend to act on your PC or your mobile device. In addition, you can find some useful tips for creating group chats on Messenger and use them to share various types of information with friends and relatives. What do you say? Are you ready to start? So let's put aside the talk and get straight into the heart of this tutorial. I wish you good reading and have fun!

Wanting to create a group on Facebook but unfamiliar with the social network founded by Mark Zuckerberg, you do not have the faintest idea how to proceed? No fear! Just follow the steps listed below to complete the task.

PC

To create a group on Facebook, from your PC, connect by the browser to the website of the social network (Of course, after you have already logged-in to your account.) and click on the green Create Group button located at the top right.

In the box that opens, type the name you want to assign to the group in the text field located under the entry 'give a name to the group', select the users you want to include in the latter by typing their names or their email addresses in the field text placed under the entry 'add some people' and then select one of the privacy settings available to decide who will be able to view the content posted. Select 'public group' to allow all users of Facebook to view the group and the content posted in it; 'group closed' to allow everyone to find the group and see the members but without being able to view the posts or 'secret group' to allow only members of the group to find it and view its posts.

After choosing the name of the group, the members that will be part of it and the privacy settings, click on the blue button 'create', select an icon that identifies the latter (e.g. the icon of the basketball, if your new group is dedicated to the world of basketball or the camera icon, if your group talks about photography) and click on the blue 'OK button'. If you wish, you can also skip this step by clicking on the 'skip' button.

Once the procedure is complete, you will be sent back to the main page of your group, which you can customize by inserting a cover image. To do so, click on the green 'upload photo' button and upload an image saved on your PC, or click on the 'select photos' button to use an image from those you have already uploaded to your Facebook profile.

To interact with friends who are part of the group, click on 'post to write a new post', click on the item 'add photo/video' to insert a multimedia content or on the voice 'live video' to start a live broadcast with other members of the group.

After creating the group, click on the options that are located immediately below the name of the group (top left); "Information", to add a description to the group; "Discussion", to share content with other members; "Members", to view members who are part of the group and to modify their roles; "Events", to create an event on the group or "Manage the group", to create scheduled posts, accept subscription requests, view the activities of administrators and so on.

To change other settings concerning the new Facebook group you have created, click the 'more' button (after clicking on the 'discussion' option) and, in the menu that appears, click on one of the options you see: "Add members", to add new members to the group or "Edit group settings", to change their name, select the group type (that is, the category related to the topic dealt with in the same), link a page, select who can publish a post and so on.

Android

To create a group on Facebook from your Android device, first, start the official social network app on your smartphone or tablet and log in to your account (if required). Then tap on the symbol (≡) located at the top right, press the 'create group' button (located under the heading Groups), type the name of the group in the text field 'give a name to this group' and, if you wish, press on the camera symbol to take or add a photo.

Then add the members to be included in the group by selecting them from the menu located at the bottom of the screen and press the 'next' button. At this point, select the privacy settings by ticking one of the available options: "Public", to create a public group and then allow anyone to view the posts and members that are part of it; "Closed", to hide the posts published on the group to those who are not members or "Private", to make the group "invisible" to those who are not part of it. After indicating the privacy options you prefer the most, press on the item 'create'. Et voilà! You've just created your group on Facebook!

Even on Android you can add new members, edit group info, create events, post photos, files and anything else using the appropriate commands that you find immediately under the name that identifies the group.

iOS

To create a group on Facebook from your iOS device, start the Facebook application, log in to your account (if necessary), press the symbol (≡) located at the bottom right and then tap on the item "Groups". In the screen that opens, press the symbol (+) located at the top right, type the name of the group in the text field "Give a name to this group" and, if you wish, press the symbol of the camera to shoot or add a photo to be used as a cover for the group.

Now, choose the members to add to the group by selecting them from the menu at the bottom and press the 'next' button. Then select the privacy settings by ticking one of the available options: "Public", to create a public group; "Closed", to hide the posts published on the group to those who are not part of it or "Private", to not allow those who are not members of the group to find it, nor to display the posts. Finally, press the item "Create" to complete the group creation procedure.

Now you can finally customize your new group, add new members, edit the info, post photos and anything else using the buttons that are immediately below the name and photo of the group itself.

Create a Chat Group Along with Your Group

If your intention is simply to converse with other Facebook users, you probably do not need to create real Facebook groups, but you can simply create simple group chats using the appropriate feature included in Messenger.

1. To create a group chat on PC, log in to your Facebook account from the browser, click on the "Messenger" icon (the lightning bolt symbol at the top), click on the item "New group", type the name of the group chat in the text field "Assign a name to the group", select the users to add to the chat and click the blue "Create" button.

2. To create a group chat on smartphones and tablets, however, start the Messenger app on your device, press on the voice "Groups", tap on the item "New group" (on Android) or "Create" (on iOS) and finally press on 'arrow symbol' (on Android) or on the "Create group entry" (on iOS) to complete the operation.

Chapter 6: How to Manage and Grow Your Facebook Group

Although it may seem a lot less "cool" than Facebook Fan Pages, a successful Facebook group can be a very interesting channel for acquiring web traffic.

In this chapter, we will discuss some very useful tips to quickly grow a Facebook group.

Before entering into the heart of this topic, I would like to make a small premise.

As you may already know, a new business begins to achieve extraordinary results when it manages to create around itself a community of similar people with shared interests.

By dealing with digital projects, our focus will be moving towards a specific type of community: the online community, the concept that is right behind a Facebook group.

Anatomy of an Online Community

Generally, an online community consists of three distinct categories of users:

1) 90% are lurkers in public;

2) 9% is the average contributors;

3) 1% is loyalists.

By lurker, we mean all those users, who hide among the members of a community, and who observe, read and use the information shared by the group, but without then going to interact and feed the conversation.

Why do they behave this way? Maybe because they believe they are not up to certain discussions, perhaps because they are shy people even in digital, or because an active participation still entails a waste of energy and time to devote to the online conversation.

Then there is the average contributor, those who together make up 9% of the community, and who share useful contents from time to time, with a variable frequency.

Lastly, there is the hardcore of the online community, 1% of the "Highlander", or rather the loyalists. They are the ones who check every day the postings spread in the group, ask questions, provide feedback to other users, and that would continue to feed the conversation in the group if it were not that at some point of the day you still have to eat and sleep!

Why Should We Focus on the Online Community of a Facebook Group?

One of the greatest benefits of developing a community on Facebook is the concept of target reach.

It is well known that a post published on a Facebook page organically reaches only a very small part of the fans of that page while, if you post the same content on a Facebook group, such sharing will be notified to every single member of the community.

So, the coverage of a post is far higher within a Facebook group and therefore also all actions related to it.

It is a very useful tool to generate web traffic for a company or personal site. But to be really effective it must be based on interesting metrics; the first of all is certainly the number of participating members.

I would, therefore, like to share some very useful tips to quickly grow the user base of a successful Facebook group.

1. Perform a preliminary analysis.

First, you need to ask yourself what the topics are, and these will be the subject of discussion of the group you want to develop.

As a result, you will be able to outline a profile of your ideal target. You cannot aim to have anyone enrolled in your Facebook community; you have to aim for the right people.

Ask yourself what your group is about. This will also be the answer concerning who really is your ideal target and what are its main characteristics.

For more profiling, you can always make sure that the members themselves define themselves better. Create market research in which you try to extrapolate more or less information about the people themselves.

2. Set goals.

Once you have a clear idea of the context on which your group revolves, it's time to think about what goal you want to pursue thanks to the Facebook group.

Do you do it for personal interests than for your brand? Why do you like networking with professionals like you? Or why do you simply want to create a community of people available to each other?

Whatever your goal, having defined it will help you adapt each situation to pursue the search results.

3. Set it up the right way.

A successful Facebook group cannot ignore, first of all, an effective name, which is as simple as it is attractive, and which can arouse curiosity and interest.

Also, think about any keywords that may be useful for searches on the internal Facebook engine.

About keywords, on the sidebar of a Facebook group or in the general settings (which you can access by clicking on the three points on the top right, near the cover image), you find a field where you can define up to 3 tags, useful for being able to be found by people.

For proper setup, take into consideration the clear description that explains what topics are treated in the group and what kind of people are registered.

In addition, you will also have to decide whether to set the group as closed, public or secret.

Starting from the last just mentioned, a secret group can be found on Facebook only by its members while a public group allows anyone, even those who are not part of the community, to see the posts shared within it.

The most effective type to quickly acquire new members in the Facebook group is the closed one.

A closed group can be found by any user during a search on Facebook using the appropriate keywords but only members can see the content published in it.

Therefore, those who are really interested in consulting what is within the group must first register. This could be an interesting profiling process though, so think about it before deciding whether to create a public or secret group.

4. Define the rules.

A healthy relationship, even the virtual one, between people must be based on clear and precise rules.

Without the guidelines, anarchy would reign, and anyone could do what they like. For example, you could only share links related to your site, constantly spamming, and you could respond in the comments using unpolished tones.

This could cause serious damage to the image of the group, difficult to repair in a short time.

In addition to this set of rules, which in digital jargon is called netiquette, you must also know how to handle cases in which heated discussions are developed between the various members.

We must, therefore, intervene, bring order and try to make sure that members can return to converse with calm tones. It is the so-called crisis management.

If, in the case, certain unpleasant situations persist, such as the continuous sharing of posts related to personal services, these articles should be removed. The responsible person should be warned and possibly banned, should he repeat this behavior several times.

5. Share useful content.

Doing Content Marketing is essential everywhere on the web, let alone in a successful Facebook group.

If you do not share valuable content, you do not feed the discussion among the members, and in that absence, there are no reasons for having to be part of an online community.

So, it is essential to design an editorial plan for your Facebook group, both in terms of content creation (creation of original content) and content curation (sharing of useful material created by other people).

In addition, if you want to optimize your work, you can always use one of the various digital tools available for post-programming on a Facebook group. For example, you could try PostPickr; it is a very useful tool that can help you take your business to the next level.

To increase the acquisition of new members and above all increase the engagement of those already present, as the group's admin, you have to publish with a certain regularity, perhaps even including questions in the texts of the posts to fuel a constructive debate between the people.

It could be once a day, twice a day, or even less, such as 2-3 times a week.

The important thing is that you understand one thing; you have to offer quality content that can really arouse interest and be useful to your subscribers.

Also, remember two other interesting features of Facebook groups:

- You can highlight a post compared to the others (the "Pin the post" feature). For example, if you find that a content can prove to be of absolute value for all members or if you want to keep the rules of the community easy to consult.

- In addition to publishing links, you can also upload files (for example, a pdf file), conduct surveys, direct live events, organize events and create documents linked to the group.

6. Shape your team.

In the beginning, when the group consists of a few tens or hundreds of members, it will not be necessary to have other collaborators to manage the community.

But when you start having thousands of participants, the situation could get out of hand; too many requests for registration links sharing of dubious usefulness, colored verbal exchanges between members, etc.

A simple remedy is to ask for help from other people, in short, set up a team of admin and moderators of the group.

Both these figures can:

- Approve or deny access to the group.

- Approve or deny the publication of a post.

- Remove posts and comments.

- Remove and ban members.

- Highlight a post.

The substantial difference that exists between an administrator of a group and a moderator is that the former can offer and remove the role of admin or moderator to other members and can make changes to the general settings of the group (name, cover photo, privacy level, etc.).

7. Promote the group.

If your tactics for organic and natural growth are tight and you want to take advantage of other techniques to quickly increase the number of members of your community, I suggest you try the following strategies.

First of all, you could start Facebook advertising campaigns to reach new people potentially interested in your group.

The Facebook Ads do not allow you to create a post natively sponsored for the promotion of a group but there is a simple trick to turn the obstacle.

So what you can do is create a Facebook page related to your group, create an advertisement linked to it and enter the URL address of your Facebook group in the field associated with the website to be sponsored.

In this way people will be conveyed directly to your online community and, if they are really interested in the topics you have proposed, they will not hesitate to register.

Another possibility to increase the number of members is represented by cross-promotions. It is about making new collaborations with other Facebook groups more or less related to the issues faced by your community. The purpose of this collaboration is based on the exchange of promotions between groups; just publish the URL address of the partner group and invite their members to follow the other group, if they consider it appropriate.

Another good way of promoting your groups is to mention the group in forums like Redditt and Quora on questions related to your niche. Just answer the questions and mention at the bottom that you have a free Facebook group that anyone can join.

Chapter 7: How to Design Your Customer Avatar

To define the profile of your ideal customer you have to remember that each individual is influenced by his position in society. Therefore, to trace the profile of the ideal customer we must answer questions related to these influences.

Social Influences

Cultural Systems and Subsystems

How old is he?

Where does he live?

Is he a man, a woman, or both of them?

What degree of education does he/she have? What schools did he/she attend?

What hobbies or passions do they have?

What are its values?

What are their beliefs?

Social class

What kind of work do they do?

How much does he or she earn?

Family

Is he or she married?

Does he or she have children? How many?

Does he lives alone and is single?

Does he live by his parents?

Marketing Influences

Do they already use a product to meet the need or solve the problem?

What kind of product do they use?

What kind of brand?

What features are relevant?

What is the benefit most appreciated?

On what price range is it oriented?

Is the high price for the customer a way of affirming their social status?

Do they associate high price with quality?

What kind of advertising influences their purchase?

Where do they usually buy?

Situational Influences

Which environments can influence the purchase (showroom, the point of sale)?

Social Environment

Within the group in which he/she lives and works, who else influences the choice of purchase?

Who uses the product?

Who pays for the product?

Are the influencer, the consumer, and the buyer the same person?

Psychological Aspects Associated with the Product

What are their fears?

What problems do they want to solve?

What consequences would it entail for an unsolved problem?

What are the challenges they're facing?

What are their wishes?

Emotional Aspects Related to the Choice of the Product

What mistakes are they afraid to commit by making a wrong choice?

What would it mean for the customer to choose a wrong product?

Now that you have answered these questions, you can trace the profile of the ideal customer.

Here's how to proceed.

Put a face to your potential client (download a photo from the internet that identifies the physical characteristics of your potential customer) and a fictitious name.

Then compile the data that will trace the characteristics of socio-demographic, psycho-graphic and consumer experiences.

This serves to have a clear representation of him/her with whom we are going to talk and with whom we want to relate.

This exercise will help you understand your customer' motivations and tailor your marketing efforts towards those motivations.

A great resource to boost this process and find people that met your avatar standards on Facebook is **lookup-id.com**. Thanks to this website, you can define your target customer through the definition of personal characteristics and get a list of people that met them.

Here is how it works.

Once that you have entered the website, go on the "extract members" section. It is very easy to find on the top right of the page. From there, you want to insert the ID of a Facebook group in your niche, which of course will contain people in target with your offer. The website will give you a complete and detailed list of the people in the group, which means that you have just discovered a goldmine, since those will be in target for what you are offering.

On this website, you can even use the FB Search function. Once you have designed the features of your typical prospect, you can insert them in this platform to get a list of users that meet those standards. It is pretty straight forward, and it is very easy to use. Our suggestion is to play around with the website and get a grasp of its potential: once you start using it, you will never get back at the classic manual research.

Now it is time to use another great tool to make life easier and start gathering a following. First of all, install **Toolkit for**

Facebook by PlugEx. What is this? It is an amazing tool that will allow you to do multiple actions in a matter of seconds. For instance, once you have your list and have launched Toolkit by PlugEx, you can then invite all those people to put a like on your page or to join your group.

When you start using these tools on a regular basis, you will be amazed at how easy it is to grow your fan base and start getting significant results. One little tip that we like to give our readers is to always use a secondary account for these operations, in order to guarantee proper privacy protection.

A practical example

Now that we have seen how to use the two software in theory, it is time to dive into the practice and use a practical example to understand the concepts better.

So, let's say that we have a shop that sells running gear and we want to find customers that are in target with the items we sell. The first thing we want to do is to go on running groups, like this one https://www.facebook.com/groups/TrailAndUltraRunning/ and look for the group id. We can do that by going here https://lookup-id.com/ and entering the previous URL. This will give us the ID of the group. After that, we just need to paste the group ID here https://lookup-id.com/get_facebookid.php to get a list of all the members that are inside the group.

After having done that, you can use Toolkit for Facebook by PlugEx to quickly invite all the people on the list you just found to leave a like on your dedicated running store page. Furthermore, you can even invite them in your very own group, where you will start your marketing process.

This is how you can use the two software together to really boost your group following.

Ref: Business Guide Offer, Instagram University

Chapter 8: How to Use ManyChat

If you're not using ManyChat, right now you're losing 60% of new potential customers.

(You're literally burning your money!)

Does this seem absurd?

Perhaps you do not know that the Chatbots are the future of marketing. For now, know that they have incredible power; to enter people's lives like never before.

Of course, there are e-mails, but, think about it. Would anyone really open a promotional e-mail?

The truth is that some when they find them in the mailbox, they almost automatically discard them. They have developed a natural tendency to delete promotional emails.

Associate the email with invasive advertising on at work or, again, at the studio. To talk to a friend, however, they use WhatsApp or Facebook Messenger.

What Do We Want to Say?

You should choose the tool to use based on your audience. If, for example, it is made up of managers, who often check e-mail, e-mail marketing may be the ideal choice.

But if perhaps, you turn to students, on average 20 years, who have much more convenience with the chat, it would be better to think of a Facebook Bot.

Do you understand what the extraordinary advantage is?

It will allow you to establish an authentic communication with your (potential or not) customer, who will open your message just like a friend. You will receive a notification on your mobile phone, you can chat with you; everything automatically.

The only thing you will have to deal with is the Bot settings.

The good news is that it is an extremely simple operation. Thanks to the services available you will use more or less 10 minutes.

However, without knowing how to use this very powerful tool effectively, you would risk undermining this work. This is why in this chapter we will introduce you to the correct configuration of a Chatbot with ManyChat.

Are you ready? Let's start!

What Is ManyChat and Create a Facebook Bot in 10 Minutes

We can say that a Bot is a program that is able to manage, in an automatic and natural way, conversations in a chat with users.

It can answer questions, offer solutions to problems and make proposals, just as if it were human.

Already from this general definition, you will have guessed that you have in your hands something potentially revolutionary for you and your business. You will free your time, increasing your results.

ManyChat, What Is It?

In short, it is the simplest and most intuitive service to create a Facebook Messenger Bot.

You do not need programming knowledge. The configuration is extremely fast, and you will immediately have the opportunity to carry out a series of actions:

- Create automated message sequences.

- Send a message to all users registered in the bot.

- Use advanced tools to increase conversions.

How ManyChat Works: The 2 Basic Tools for Bot Marketing

We have just said that ManyChat is the easiest to use Messenger Marketing tool.

Despite its simplicity, however, it offers a number of features that make it complete and effective for your web marketing strategy.

Do you want some examples?

- Automatic sequences.

This is the series of messages that, automatically, ManyChat will send to the user. You can set them as you like, based on your lead generation strategy.

- Growth Tools.

Here, this is the real bomb among the tools offered. They are a series of "extensions" that add functionality to ManyChat. The most famous is Facebook Comments Tool, which allows you to convert whoever who comments on a particular post into a member of the Bot.

(These are just some of the features you'll have available.)

Well, to make you better understand the functioning of a Chatbot; let's assume a case of real use:

The user Luke comments on the post of your product with the specific keyword that you have set. Your bot turns on and automatically sends your opt-in message; basically, a welcome message to confirm the user's interest.

Subsequently, based on the behavior of Luke, the bot will send him different messages, to achieve the goals you have set. These messages are part of an "automated sequence" that you created earlier.

The important thing is to never be intrusive - keep this in mind during the setup phase. Each time you send a message, a notification will be sent to Luke's mobile phone. This means two things:

- You will enter his daily life, like never before (remember what we said at the beginning of the chapter?).

- You will have to manage this opportunity in the best way, so as not to frustrate him (most people are not accustomed to this tool just yet, so do not overuse it).

Conclusion

Thanks for making it through to the end of **Facebook Marketing Strategies: Zero Cost Facebook Marketing Plan for Small Business**, let's hope it was informative and able to provide you with all of the tools you need to achieve your goals whatever it is that they may be. Just because you've finished this book doesn't mean there is nothing left to learn on the topic. Expanding your horizons is the only way to find the mastery you seek.

The next step is to stop reading and to get started doing whatever it is that you need to do in order to ensure that those you care about will be properly taken care of should the need arise. If you find that you still need help getting started, you will likely have better results by creating a schedule that you hope to follow including strict deadlines for various parts of the tasks as well as the overall completion of your preparations.

Studies show that complex tasks that are broken down into individual pieces, including individual deadlines, have a much greater chance of being completed when compared to something that has a general need of being completed but no real timetable for doing so. Even if it seems silly, go ahead and set your own deadlines for completion, complete with indicators of success and failure. After you have successfully completed all of your required preparations you will be glad you did.

Once you have finished your initial preparations it is important to understand that they are just that, only part of a larger plan of preparation. Your best chances for overall success will come by taking the time to learn as many marketing skills as possible. Only by using your prepared status as a springboard to greater preparation will you be able to truly rest soundly knowing that you are prepared for anything and everything that the market decides to throw at you.

Finally, if you found this book useful in any way, a review on Amazon is always appreciated!